Harvest Season

by Helen...

Consultant:
Adria F. Klein, PhD
California State University, San Bernardino

CAPSTONE PRESS
a capstone imprint

Wonder Readers are published by Capstone Press,
1710 Roe Crest Drive, North Mankato, Minnesota 56003.
www.capstonepub.com

Copyright © 2013 by Capstone Press, a Capstone imprint. All rights reserved.
No part of this publication may be reproduced in whole or in part, or stored in a retrieval system, or transmitted in any form or by any means, electronic, mechanical, photocopying, recording, or otherwise, without written permission of the publisher. For information regarding permission, *write* to Capstone Press, 1710 Roe Crest Drive, North Mankato, Minnesota 56003.

Library of Congress Cataloging-in-Publication Data
Gregory, Helen.
  Harvest season / Helen Gregory.—1st ed.
      p. cm.
  Includes index.
  ISBN 978-1-4765-0042-3 (library binding)
  ISBN 978-1-4296-8058-5 (paperback)
  ISBN 978-1-4765-0854-2 (eBook PDF)
  1. Harvesting—Juvenile literature. I. Title.
  SB129.G74 2013
  631.5'5—dc23   2011023942

Summary: Describes a variety of foods that are harvested in the fall.

Editorial Credits
Maryellen Gregoire, project director; Mary Lindeen, consulting editor; Gene Bentdahl, designer; Sarah Schuette, editor; Wanda Winch, media researcher; Eric Manske, production specialist

Photo Credits
Capstone Studio: Gary Sundermeyer, 8, 9, Karon Dubke, cover, 1, 5, 13; Shutterstock: Denis and Yulia Pogostins, 6, Elena Elisseeva, 18, Fotokostic, 16, Kletr, 4, mycola, 14, Neil Roy Johnson, 11, Orientaly, 15, Shestakoff, 7, smart.art, 17, Smileus, 12, Tish1, 10

Word Count: **201**   Guided Reading Level: **I**   Early Intervention Level: **14**

Printed in China.
092012   006934LEOS13

Table of Contents

Harvest Time 4
Fruit 6
Vegetables 10
Grains 14
Now Try This! 19
Glossary 19
Internet Sites 20
Index 20

Note to Parents and Teachers

The Wonder Readers Next Steps: Social Studies series supports national social studies standards. These titles use text structures that support early readers, specifically with a close photo/text match and glossary. Each book is perfectly leveled to support the reader at the right reading level, and the topics are of high interest. Early readers will gain success when they are presented with a book that is of interest to them and is written at the appropriate level.

Harvest Time

Harvesting is gathering crops from the fields. The harvest marks the end of the growing season.

Harvesting is doing what it takes to get the food to **market**. Sometimes that means sorting, cleaning, and packing the crop.

Fruit

These strawberries are ready to be picked. It is time to harvest them.

You can pick the strawberries from the strawberry plant.

The apples are ready to be picked.
It is time to harvest them.

You can pick apples from a tree by hand.

Vegetables

Potatoes grow underground. These potatoes are ready to be dug up.

It is time to harvest.
You need to dig the potatoes
and sort them.

Pumpkins grow on **vines**.
These pumpkins are ready
to be picked.

It is time to harvest.
Finding a perfect pumpkin
is sometimes very hard.

Grains

Wheat is a **grain**.
It grows on tall stems.

It is time to harvest.
The wheat is ready to be cut.
Machines cut and sort the wheat.

Corn is a grain.

It grows on a **stalk** in a field.

This corn is ready to be picked.

It is time to harvest.
Machines pick and shell the corn.

The harvest is done.
Now the crops are ready
to go to the market.

Now Try This!

Make a list of the things you ate for breakfast or lunch today. Separate each item according the food group it belongs to. What are the similarities between the foods mentioned in this book and the foods you included on your list? How do you think the foods you ate were grown and harvested?

Glossary

grain the seed of a cereal plant such as wheat, rice, corn, rye, or barley

market a place where people can buy food and other items

stalk a stem of a plant; the stalk supports or connects parts of a plant

vine a plant with a long, thin stem that climbs trees, fences, or other supports

Internet Sites

FactHound offers a safe, fun way to find Internet sites related to this book. All of the sites on FactHound have been researched by our staff.

Here's all you do:

Visit *www.facthound.com*

Type in this code: 9781476500423

Super-cool stuff! Check out projects, games and lots more at www.capstonekids.com

Index

apples, 8–9
corn, 16–17
crops, 4–5, 18
growing seasons, 4
markets, 5, 18
potatoes, 10–11
pumpkins, 12–13
strawberries, 6–7
wheat, 14–15